天王星

侧向自转的星球

URANUS

The Sideways-Spinning Planet

（英国）埃伦·劳伦斯/著　张　骁/译

江苏凤凰美术出版社

著作权合同登记图字：10-2022-144

图书在版编目（CIP）数据

天王星：侧向自转的星球 /（英）埃伦·劳伦斯著；

张骁译 . -- 南京：江苏凤凰美术出版社，2025. 1.

（环游太空）. -- ISBN 978-7-5741-2027-3

Ⅰ . P185.6-49

中国国家版本馆 CIP 数据核字第 2024H9727V 号

策　　　　划	朱　婧
责 任 编 辑	高　静　奚　鑫
责 任 校 对	王　璇
责任设计编辑	樊旭颖
责 任 监 印	生　嫄
英 文 朗 读	C.A.Scully
项 目 协 助	邵楚楚　乔一文雯

丛 　书　 名	环游太空
书　　　　名	天王星：侧向自转的星球
著　　　　者	（英国）埃伦·劳伦斯
译　　　　者	张　骁
出 版 发 行	江苏凤凰美术出版社（南京市湖南路 1 号 邮编：210009）
印　　　刷	南京新世纪联盟印务有限公司
开　　　本	710 mm×1000 mm　1/16
总 印 张	18
版　　　次	2025 年 1 月第 1 版
印　　　次	2025 年 1 月第 1 次印刷
标 准 书 号	ISBN 978-7-5741-2027-3
总 定 价	198.00 元（全 12 册）

版权所有　侵权必究

营销部电话：025-68155675　营销部地址：南京市湖南路 1 号

江苏凤凰美术出版社图书凡印装错误可向承印厂调换

目录 Contents

书中加粗的词语见词汇表解释。

Words shown in **bold** in the text are explained in the glossary.

欢迎来到天王星
Welcome to Uranus

想象一下，你正在飞往一个离地球20多亿千米的地方。

Imagine flying to a world that is over 2 billion kilometers from Earth.

随着飞船越来越近，在黑暗的宇宙中你看到了一个美丽的蓝色星球。

As your spacecraft gets near, you see a beautiful, blue ball in the darkness of space.

你飞得越来越近，潜入到旋转的气体和云层中。

You fly much closer and dive into swirling **gases** and clouds.

然而在这个遥远的世界里，你发现根本没地方着陆。

You will find nowhere solid to land, though,on this faraway world.

因为这颗巨大的蓝色球体全都是由极寒的气体和液体构成的。

That's because the huge, blue ball is made only of icy gases and liquids.

欢迎来到行星天王星！

Welcome to the **planet** Uranus!

人类从未到达过天王星，但是有一个太空飞行器做到了。198 "旅行者2号"成为第一个也是唯一一个飞跃天王星的空间探测器

No humans have ever visited Uranus, but a spacecraft has. In 1986, V 2 became the first and only space **probe** to fly past Uranus.

这张图片是电脑合成的，它显示了在近距离观察下天王星的模样。
This picture was created on a computer. It shows how it might look to fly very close to Uranus.

太阳系 The Solar System

天王星以超过24 000千米每小时的速度在太空移动。

Uranus is moving through space at over 24,000 kilometers per hour.

它围绕着太阳做一个巨大的圆周运动。

It is moving in a huge circle around the Sun.

天王星是围绕太阳公转的八大行星之一。

Uranus is one of eight planets circling the Sun.

八大行星分别是水星、金星、我们的母星地球、火星、木星、土星、天王星和海王星。

The planets are called Mercury, Venus, our home planet Earth, Mars, Jupiter, Saturn, Uranus, and Neptune.

冰冻的彗星和被称为"小行星"的大型岩石也围绕着太阳公转。

Icy **comets** and large rocks, called **asteroids**, are also moving around the Sun.

太阳、行星和其他天体共同组成了"太阳系"。

Together, the Sun, the planets, and other space objects are called the **solar system**.

太阳系中的大多数小行星都集中在被称为"小行星带"的环状带中。

Most of the asteroids in the solar system are in a ring called the asteroid belt.

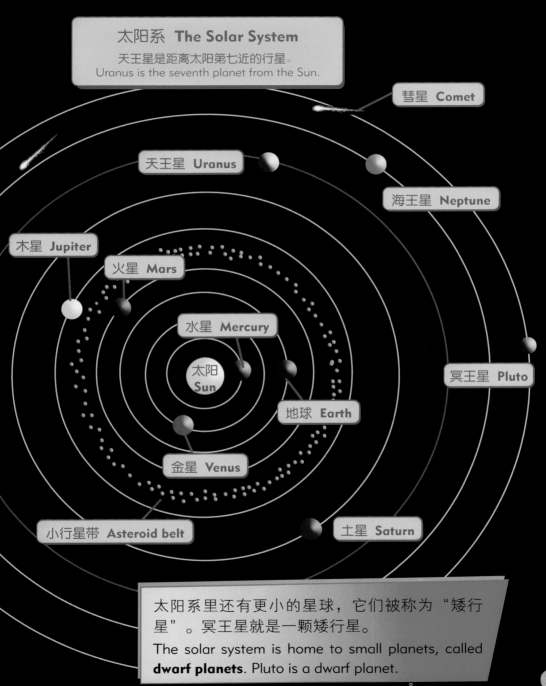

太阳系 **The Solar System**
天王星是距离太阳第七近的行星。
Uranus is the seventh planet from the Sun.

彗星 **Comet**

天王星 **Uranus**

海王星 **Neptune**

木星 **Jupiter**

火星 **Mars**

水星 **Mercury**

太阳
Sun

冥王星 **Pluto**

地球 **Earth**

金星 **Venus**

小行星带 **Asteroid belt**

土星 **Saturn**

太阳系里还有更小的星球，它们被称为"矮行星"。冥王星就是一颗矮行星。

The solar system is home to small planets, called **dwarf planets**. Pluto is a dwarf planet.

天王星的奇幻之旅
Uranus's Amazing Journey

行星围绕太阳公转一圈所需的时间被称为"一年"。

地球绕太阳公转一圈需要略多于365天的时间，所以地球上的一年有365天。

天王星比地球离太阳更远，所以它绕太阳公转的路程更长。

天王星绕太阳公转一圈大约需要84个地球年。

你能想象在天王星上要等84个地球年才能过一个生日吗？

The time it takes a planet to **orbit**, or circle, the Sun once is called its year.

Earth takes just over 365 days to orbit the Sun, so a year on Earth lasts 365 days.

Uranus is farther from the Sun than Earth, so it must make a much longer journey.

It takes Uranus just over 84 Earth years to orbit the Sun.

Can you imagine waiting 84 Earth years to celebrate your first birthday in Uranus years?

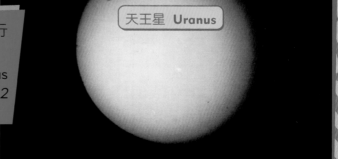

这张天王星的照片是由"旅行者2号"空间探测器拍摄的。
This photograph of Uranus was taken by the *Voyager 2* space probe.

天王星 Uranus

地球绕太阳公转一圈的路程约为9.4亿千米，而天王星绕太阳公转一圈的距离约为180亿千米！

To orbit the Sun once, Earth makes a journey of about 940 million km. Uranus must make a journey of about 18 billion km!

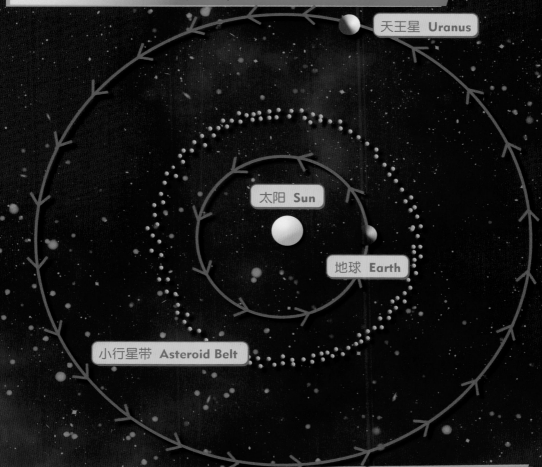

天王星 Uranus

太阳 Sun

地球 Earth

小行星带 Asteroid Belt

在这张图片上，我们去掉了太阳系中除了地球和天王星以外的其他行星。现在可以看出天王星绕太阳公转的超长旅程了吧！

In this picture, we've taken away all the planets in the solar system except for Earth and Uranus. Now it's easy to see Uranus's super-long journey around the Sun!

近距离观察天王星
A Closer Look at Uranus

天王星是太阳系中第三大行星，仅次于木星和土星。

Uranus is the third-largest planet in the solar system, behind Jupiter and Saturn.

不同于地球是一颗岩质行星，天王星没有固体表面。

Unlike Earth, which is a rocky planet, Uranus doesn't have a solid surface.

天王星的外围覆盖着一层气体和云层，称为"大气层"。

The planet has an outer layer of gases and clouds called an **atmosphere**.

在大气层里，风速可以达到900千米每小时。

Inside the atmosphere, winds blow at up to 900 km/h.

在大气层下面，天王星是一个由极寒液体组成的巨大球体。

Beneath its atmosphere, Uranus is a huge ball of icy liquids.

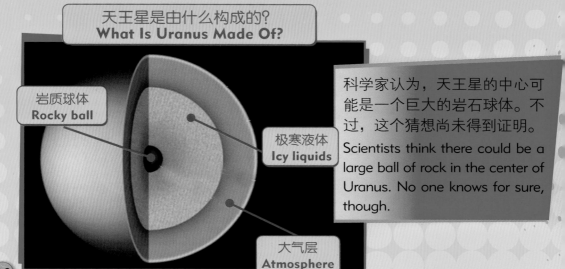

天王星是由什么构成的?
What Is Uranus Made Of?

岩质球体
Rocky ball

极寒液体
Icy liquids

大气层
Atmosphere

科学家认为，天王星的中心可能是一个巨大的岩石球体。不过，这个猜想尚未得到证明。

Scientists think there could be a large ball of rock in the center of Uranus. No one knows for sure, though.

地球 **Earth**

天王星 **Uranus**

天王星的直径大约是我们的母星——地球的4倍。

Uranus is nearly four times wider than our home planet, Earth.

11

天王星是如何自转的
How Uranus Spins

绕着太阳公转的行星也在自转，就像陀螺一样。

As a planet orbits the Sun, it also spins, or **rotates**, like a top.

当它们自转的时候，大都围绕一条近乎竖直的轴转。

As they rotate, most planets are in a nearly upright position.

但是天王星却是个例外，它竟是"躺"着旋转的！

Uranus, however, spins on its side!

有些科学家认为天王星以前也是竖直旋转的。

Some scientists think that Uranus once rotated in an upright position.

只是后来，它遭到了其他巨大的天体的撞击。

Then, it was hit by another huge space object.

这次撞击把天王星撞歪了，于是从此以后它就变成了一颗侧向自转的行星！

This **collision** knocked Uranus onto its side, and turned it into a sideways-spinning planet!

木星 Jupiter

这张图片显示的是木星，它就是一颗竖直自转的行星。

This picture shows the planet Jupiter. Jupiter rotates in an upright position.

地球 Earth

地球在旋转的时候会轻微地向一侧倾斜。

Earth is slightly tilted to one side as it spins.

天王星 Uranus

天王星是侧着旋转的，它的自转方向也和大多数行星相反。看看图片上环绕着天王星、地球和木星的箭头，你就能观察到这一点。

Uranus spins on its side. It also spins in the opposite direction than most other planets. You can see this by looking at the arrows around Uranus, Earth, and Jupiter.

天王星的卫星家族
A Family of Moons

天王星有一个庞大的卫星家族绕着它旋转。

Uranus has a large family of small worlds orbiting around it.

这些温度极低的岩质天体都是这颗行星的卫星。

These icy, rocky space objects are the planet's **moons**.

我们的家园——地球，只有一颗卫星。

Earth, our home planet, has just one moon.

而天王星至少有28颗卫星，而且科学家们认为也许还会有更多！

Uranus has at least 28 moons, and scientists think there may be more!

天王星最小的那些卫星中，有不少直径只有16千米左右。

Many of Uranus's smallest moons are only about 16 km wide.

它们在天王星面前可以说是非常渺小，因为天王星的直径差不多有50 000千米！

That's tiny compared to huge Uranus, which is about 50,000 km wide!

艾瑞尔（天卫一）Ariel

这张特写图片展现了这颗名为艾瑞尔的卫星的表面。

This close-up picture shows the surface of a moon named Ariel (AIR-ee-uhl).

地球 Earth

地球的卫星
Earth's moon

泰坦尼亚（天卫三）
Titania

天王星最大的卫星名为泰坦尼亚。它的直径大约为1 600千米，大小约是地球卫星的一半。

Uranus's largest moon is named Titania (ty-TANE-ee-yuh). It is nearly 1,600 km wide. That's about half the size of Earth's moon.

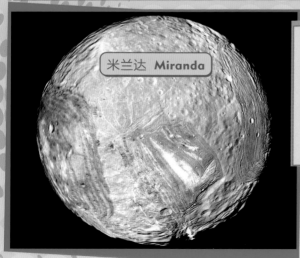

米兰达 Miranda

天王星的卫星米兰达表面有巨大的峡谷，深度大约是科罗拉多大峡谷的12倍！

Uranus's moon Miranda has giant **canyons** on its surface. The canyons are 12 times as deep as the Grand Canyon!

惊人的大发现
A Surprising Discovery

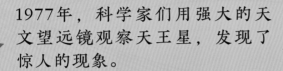

1977年，科学家们用强大的天文望远镜观察天王星，发现了惊人的现象。

他们发现竟然有圆环围绕在这颗行星周围。

以前可从来没有人看到过这些行星环。

其他望远镜和"旅行者2号"探测器帮助我们得到了更多有关天王星环的知识。

现在我们知道在天王星周围有至少13个微弱的圆环。

In 1977, scientists watching Uranus through powerful telescopes saw something surprising.

They discovered that there are rings circling the planet.

No one had ever seen the rings before.

Other telescopes and the *Voyager 2* probe have helped us learn more about the rings.

We now know that there are at least 13 faint rings circling Uranus.

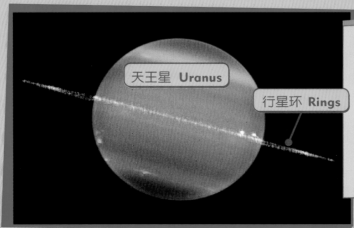

天王星 Uranus

行星环 Rings

这张照片显示了遥远的天王星和它的行星环。照片是由夏威夷一座山上的望远镜拍摄的。

This is a photo of faraway Uranus and its rings. It was taken by a telescope on a mountain in Hawaii.

天王星环是由大量冰冷、黑暗的物质组成的。
不过还没有人知道这些物质究竟是什么。
Uranus's rings are made of billions of pieces of
icy, dark material. No one knows for sure what
the material is, though.

天王星 Uranus

行星环上冰冷、黑暗的碎片
Icy, dark pieces in rings

这张图片是电脑合成的，它显
示了在近距离观察下行星环的
模样。
This picture was created on a
computer. It shows how the rings
might look up close.

探测天王星的任务
A Mission to Uranus

1977年8月，空间探测器"旅行者2号"从地球发射。

In August 1977, the space probe *Voyager 2* blasted off from Earth.

1986年1月，它抵达天王星，并近距离地飞越了天王星。

In January 1986, it reached Uranus and did a close **flyby**.

"旅行者2号"和天王星的会晤只持续了5个半小时。

Voyager 2's meeting with Uranus lasted just five and a half hours.

在这段时间里，探测器发现了10颗新卫星和两个新的天王星环。

During that time, the probe found 10 new moons and two new rings circling Uranus.

它还发现这颗行星上有一个充满沸水的大洋！

It also discovered an ocean of boiling water on the planet!

至今为止，还没有别的空间探测器来过天王星！

No other space probe has visited Uranus—yet!

火箭 Rocket

"旅行者2号"搭载火箭从地球上发射。
Voyager 2 blasts off from Earth aboard a rocket.

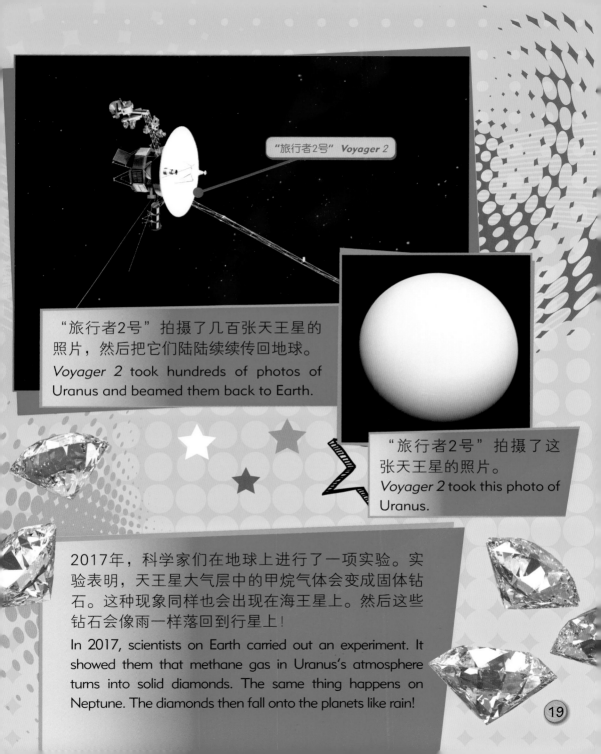

"旅行者2号" *Voyager 2*

"旅行者2号"拍摄了几百张天王星的照片，然后把它们陆陆续续传回地球。
Voyager 2 took hundreds of photos of Uranus and beamed them back to Earth.

"旅行者2号"拍摄了这张天王星的照片。
Voyager 2 took this photo of Uranus.

2017年，科学家们在地球上进行了一项实验。实验表明，天王星大气层中的甲烷气体会变成固体钻石。这种现象同样也会出现在海王星上。然后这些钻石会像雨一样落回到行星上！

In 2017, scientists on Earth carried out an experiment. It showed them that methane gas in Uranus's atmosphere turns into solid diamonds. The same thing happens on Neptune. The diamonds then fall onto the planets like rain!

有趣的天王星知识
Uranus Fact File

以下是一些有趣的天王星知识：天王星是距离太阳第七近的行星。

Here are some key facts about Uranus, the seventh planet from the Sun.

天王星的发现
Discovery of Uranus

天王星在1781年被威廉·赫歇尔发现。尽管也有其他人曾用望远镜观察到天王星，但是赫歇尔却是第一个确认它为行星的人。

Uranus was discovered in 1781 by William Herschel. Other people had seen Uranus through telescopes, but Herschel was the first to say for sure that it was a planet.

天王星是如何得名的
How Uranus got its name

这颗行星以古希腊神话中掌管天空的神命名。

The planet is named after the Greek god of the sky.

行星的大小
Planet sizes

这张图显示了太阳系八大行星的体积对比。

This picture shows the sizes of the solar system's planets compared to each other.

水星 Mercury

地球 Earth

木星 Jupiter

天王星 Uranus

太阳 Sun

火星 Mars

金星 Venus

土星 Saturn

海王星 Neptune

天王星的大小
Uranus's size

天王星的直径约50 724千米

About 50,724 km across

天王星自转一圈需要多长时间
How long it takes for Uranus to rotate once

大约17个地球时

About 17 Earth hours

天王星与太阳的距离
Uranus's distance from the Sun

天王星与太阳的最短距离是2 734 998 229千米。
天王星与太阳的最远距离是3 006 318 143千米。

The closest Uranus gets to the Sun is 2,734,998,229 km.
The farthest Uranus gets from the Sun is 3,006,318,143 km.

天王星绕太阳轨道的长度
Length of Uranus's orbit around the Sun

18 026 802 831千米
18,026,802,831 km

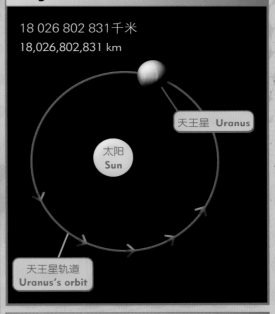

天王星 Uranus

太阳
Sun

天王星轨道
Uranus's orbit

天王星围绕太阳公转的平均速度
Average speed at which Uranus orbits the Sun

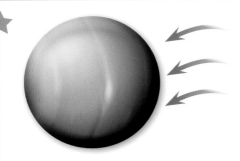

每小时24 477千米
24,477 km/h

天王星上的一年
Length of a year on Uranus

大约30 687个地球天
（大约84个地球年）

30,687 Earth days
(about 84 Earth years)

天王星的卫星
Uranus's Moons

天王星至少有27颗已知的卫星。还有更多的有待发现。

Uranus has at least 27 known moons. There are possibly more to be discovered.

天王星上的温度
Temperature on Uranus

零下216摄氏度

-216°C

动动手吧：宇宙飞行棋
Get Crafty : Zoom Into Space Game

制作一副你自己的桌游，试试飞越太阳系吧！

在一张正方形纸板上画出25个方格。在格子里分别画上行星、卫星、太阳、小行星和彗星。现在制订规则，然后和朋友们一起飞向浩瀚的宇宙吧！

游戏创意

这里有些点子可以帮你创造自己的规则：

- 试着扔骰子，扔到几就走几格。想想看，如果你降落在某个天体的图片上，会发生什么事呢？
- 或许降落在太阳上可以得10分，降落在行星上可以得5分。
- 也许降落在小行星上就要退后两格。

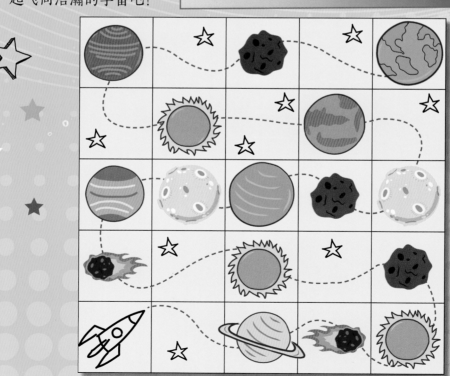

词汇表 Glossary

小行星 | **asteroid**

围绕太阳公转的大块岩石，有些小得像辆汽车，有些大得像座山。

大气层 | **atmosphere**

行星、卫星或恒星周围的一层气体。

峡谷 | **canyon**

陡峭的深谷。

撞击 | **collision**

两个物体冲撞到一起。

彗星 | **comet**

由冰、岩石和尘埃组成的天体，围绕太阳公转。

矮行星 | **dwarf planet**

围绕太阳运行的圆形或近圆形天体，比八大行星小得多。

飞越 | **flyby**

航天器近距离飞过行星、月球或其他天体的行为。飞越过程中，航天器会靠近某一行星，以便对它进行仔细的研究，然后把信息传回地球。

气体 | gas

无固定形状或大小的物质，如氧气或氦气。

卫星 | moon

围绕行星运行的天体。通常由岩石或岩石和冰构成。直径从几千米到几百千米不等。地球有一个卫星，名为"月球"。

公转 | orbit

围绕另一个天体运行。

行星 | planet

围绕太阳公转的大型天体：一些行星，如地球，主要是由岩石组成的；其他的行星，如木星，主要是由气体和液体组成的。

探测器 | probe

不载人太空飞船。通常被送往行星或其他天体，用于拍摄照片并收集信息，由地球上的科学家操作控制。

自转 | rotate

物体自行旋转的运动。

太阳系 | solar system

太阳和围绕太阳公转的所有天体，如行星及其卫星、小行星和彗星。